MY BOOK OF FIRST 500 WORDS

This book belongs to

Alphabets

Aa

Aeroplane

Bb

Bus

Cc

Car

Dd

Dice

Ee

Elephant

Ff

Fish

Gg

Goat

Hh

Hen

Ii

Ice Cream

Jj

Jet

Kk

Kite

Ll

Lemon

Mm

Mushroom

Nn

Nest

Oo

Owl

Pp

Parrot

Qq

Queue

Rr

Rose

Ss

Ship

Tt

Train

Uu

Umbrella

Vv

Vegetables

Ww

Watch

Xx

X Mas Tree

Yy

Yacht

Zz

Zip

Colours

Orange

Green

Pink

Red

Violet

Grey

Yellow

Blue

Black

White

Brown

Shapes

Star Starfish

Diamond Kite

Heart Leaf

Crescent Banana

Oval Egg

Rectangle Book

Triangle Sandwich

Circle Football

Square Dice

Cone Birthday Cap

Numbers

1
One
2
Two
3
Three
4
Four
5
Five
6
Six
7
Seven
8
Eight
9
Nine
10
Ten

11
Eleven
12
Twelve
13
Thirteen
14
Fourteen
15
Fifteen
16
Sixteen
17
Seventeen
18
Eighteen
19
Nineteen
20
Twenty

Opposites

New

Old

Hot

Cold

Slow

Fast

Sour

Sweet

Small

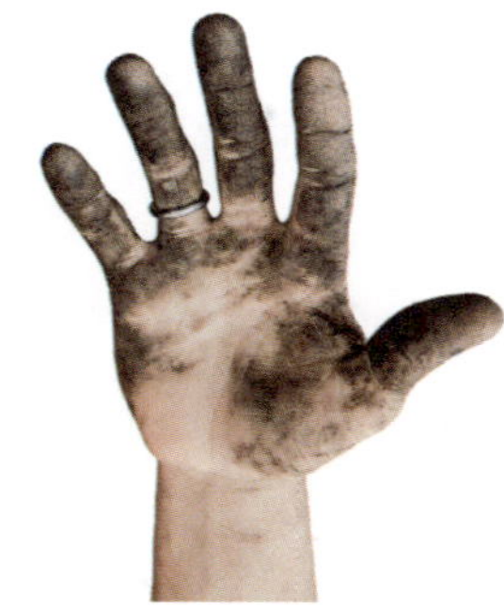

Big

Clean

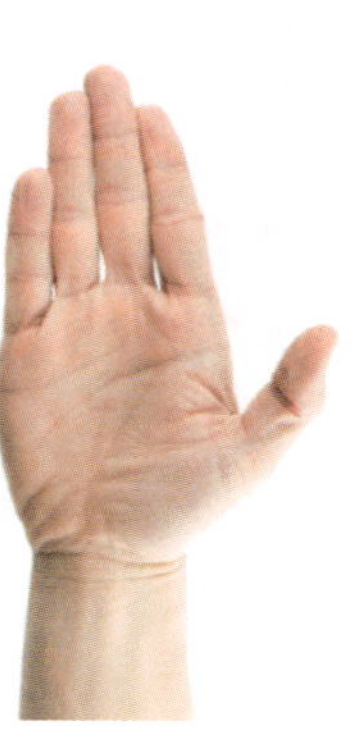

Dirty

Open

Close

White

Black

Happy

Sad

Sit

Stand

Back

Front

Soft

Hard

Vegetables

Capsicum
Broccoli
Carrot
Onion
Bitter Gourd
Coriander
Pumpkin
Beetroot
Cabbage
Tomato

Cucumber
Cauliflower
Ginger
Green Chilli
Eggplant
Spinach
Lady Finger
Turnip
Potato
Green Pea

Fruits

Plum
Mango
Kiwi
Strawberry
Apple
Pomegranate
Custard Apple
Star Fruit
Orange
Pear

Blackberry

Papaya

Pineapple

Grapes

Guava

Cherry

Peach

Banana

Muskmelon

Watermelon

Land Transport

Lorry

Tow Truck

Fire Truck

Tractor

Taxi

Oil Tanker

Snowmobile

Scooter

Go Cart

Caravan

Cycle

Jeep

Air Transport

Aeroplane

Blimp

Helicopter

Parachute

Hot Air Balloon

Rocket

Sea Aeroplane

Jet

Water Transport

Food

Eggs
Cookies
Muffin
Chicken
Hotdog
Pizza
Burger
Soup
Cornflakes
Fish

Omelette

Yogurt

Milk

Bread

Sausage

Cheese

Sandwich

Rice

Ice Cream

Fruit Salad

Birds

Vulture
Flamingo
Eagle
Crow
Peacock
Parrot
Sparrow
Duck
Crane
Robin

Pigeon
Owl
Ostrich
Nightingale
Goose
Dove
Woodpecker
Penguin
Swan
Kingfisher

Pet Animals

Farm Animals

Wild Animals

Deer
Chimpanzee
Lion
Gorilla
Panda
Cheetah
Zebra
Boar
Hare
Giraffe
Polar Bear
Monkey

Sea Animals

Whale
Stingray
Clownfish
Sea Turtle
Catfish
Dragonfish
Shrimp
Otter
Starfish
Shells

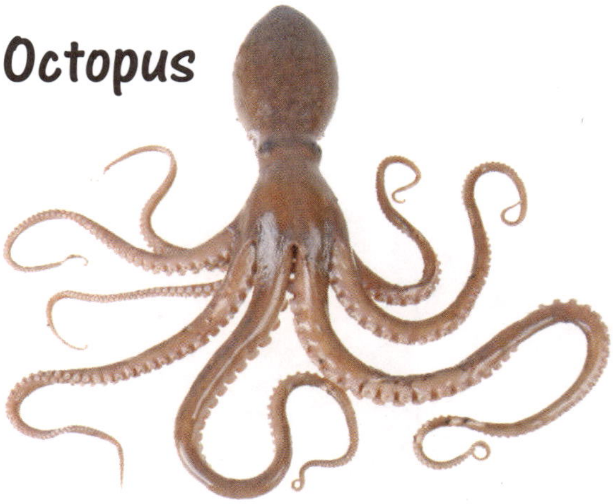

Octopus

Walrus

Squid

Shark

Sea Horse

Dolphin

Eel

Crab

Jellyfish

Coral

Seal

Lobster

Baby Animals

Fawn
Cygnet
Owlet
Bear Cub
Colt
Kitten
Kid
Joey
Polar Bear Cub
Tiger Cub
Hatchling
Baby Platypus

People at Work

Teacher

Astronaut

Chef

Electrician

Architect

Firefighter

Pilot

Doctor

Baker

Scientist

Photographer

Plumber

Nurse

Gardener

Police Officer

Soldier

Farmer

Reporter

Toys

Ball
Robot
Toy Cycle
Doctor Set
Rattle
Doll
Duck
Drum
Ring Donut
Toy Train

Piano

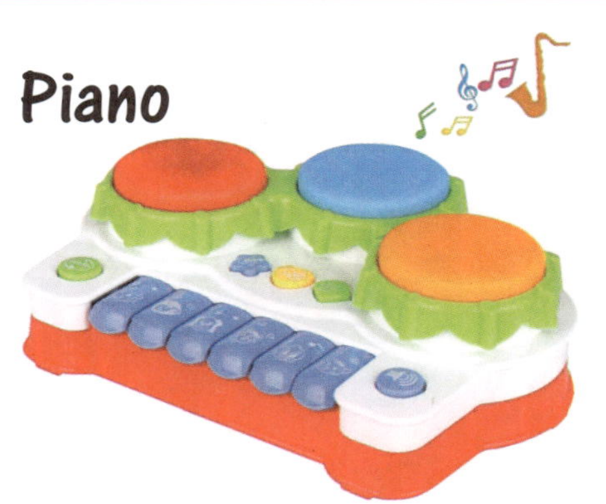

Blocks

Kitchen Set

Car

Bike

Teddy

Rocking Horse

Activity Toy

Remote Car

Police Car

Objects Around the Baby

Pillow
Food
Spoon
Potty Seat
Booties
FRECHE FREUNDE
Bowl
Bib
Pacifier
Crib
Play Mat
Socks
Bathtub

In the Living Room

In the Bedroom

Dressing Table

Drawer

Wardrobe

Table Lamp

Pillow

Fan

Air Conditioner

Doormat

Bed

Quilt

In the Bathroom

Hair Dryer
Loofah
Mirror
FAMILY
Exhaust Fan
Tissue Paper
Mug
Towel
Toothbrush
MOUTH
FRESHENER
HAIR
OIL
Shampoo
Mouth Freshener
Hair Oil
Laundry Hamper
Shampoo

Kitchen Appliances

Chimney
Refrigerator
Gas Lighter
Plate
Washbasin
Spatula
Tray
Jar
Glass
Whisk
Chopping Board

In the School

Pen

Marker

Ruler

Pencil

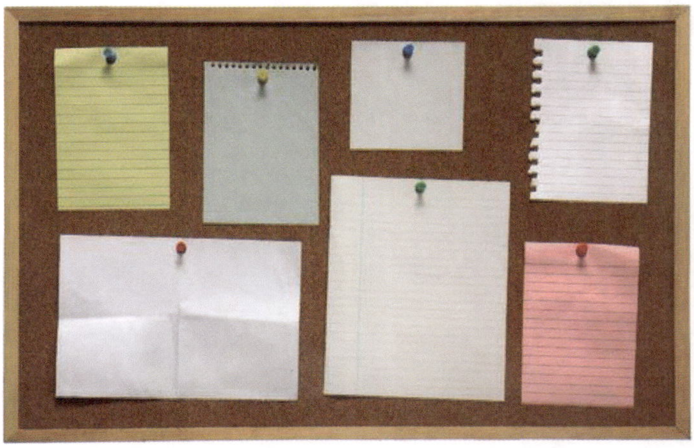

Notice Board

Board

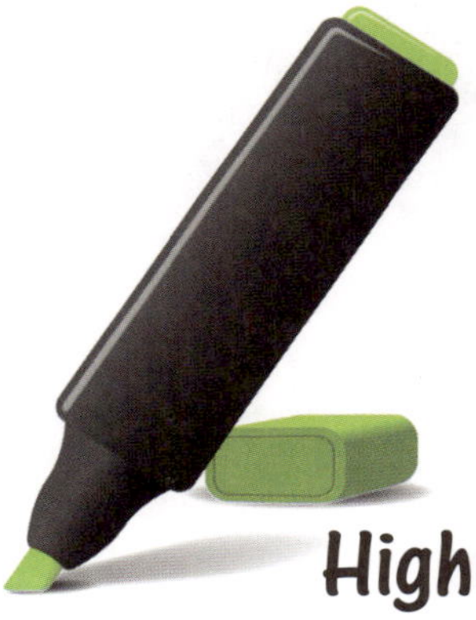

Highlighter

Pen Holder

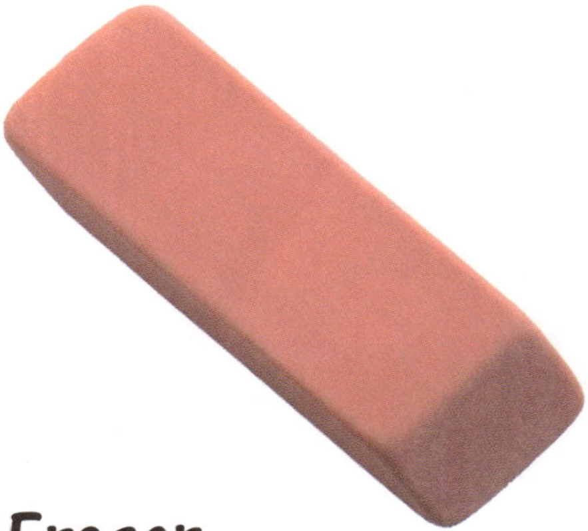

Eraser

Colours

School Bag

In the Garden

Green House

Honeycomb

Garden Hose

Leaves

Grass

Nest

Clothes

Raincoat

Cardigan

T-Shirt

Long Coat

Nightgown

Jeans

Jacket

Sweatshirt

Socks

Sports

Cricket

Volleyball

Skiing

Weight Lifting

Archery

Golf

Table Tennis

Boxing

Rugby
Soccer
Karate
Baseball
Tennis
Horse Racing
Basketball
Cycling

Flowers

Sunflower
Pansy
Lily
Dahlia
Petunia
Hibiscus
Orchid
Carnation
Daffodil

Tulip

Periwinkle

Rose

Aster

Jasmine

Lotus

Water Lily

Canna

Marigold

Bluebell

Accessories

My Senses

Action Words

Praying

Dancing

Picking

Smiling

Eating

Thinking

Feeling Sad

Jumping

Playing

Pulling

Pushing

Hiding

Listening

Drinking

Laughing

Reading

Action Words

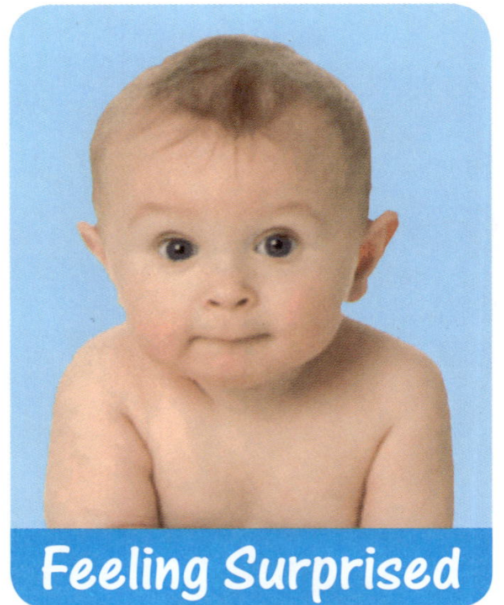
Feeling Surprised

Sleeping

Crying

Frowning

Yawning

Feeling Excited

Talking

Bending

Writing

My Face

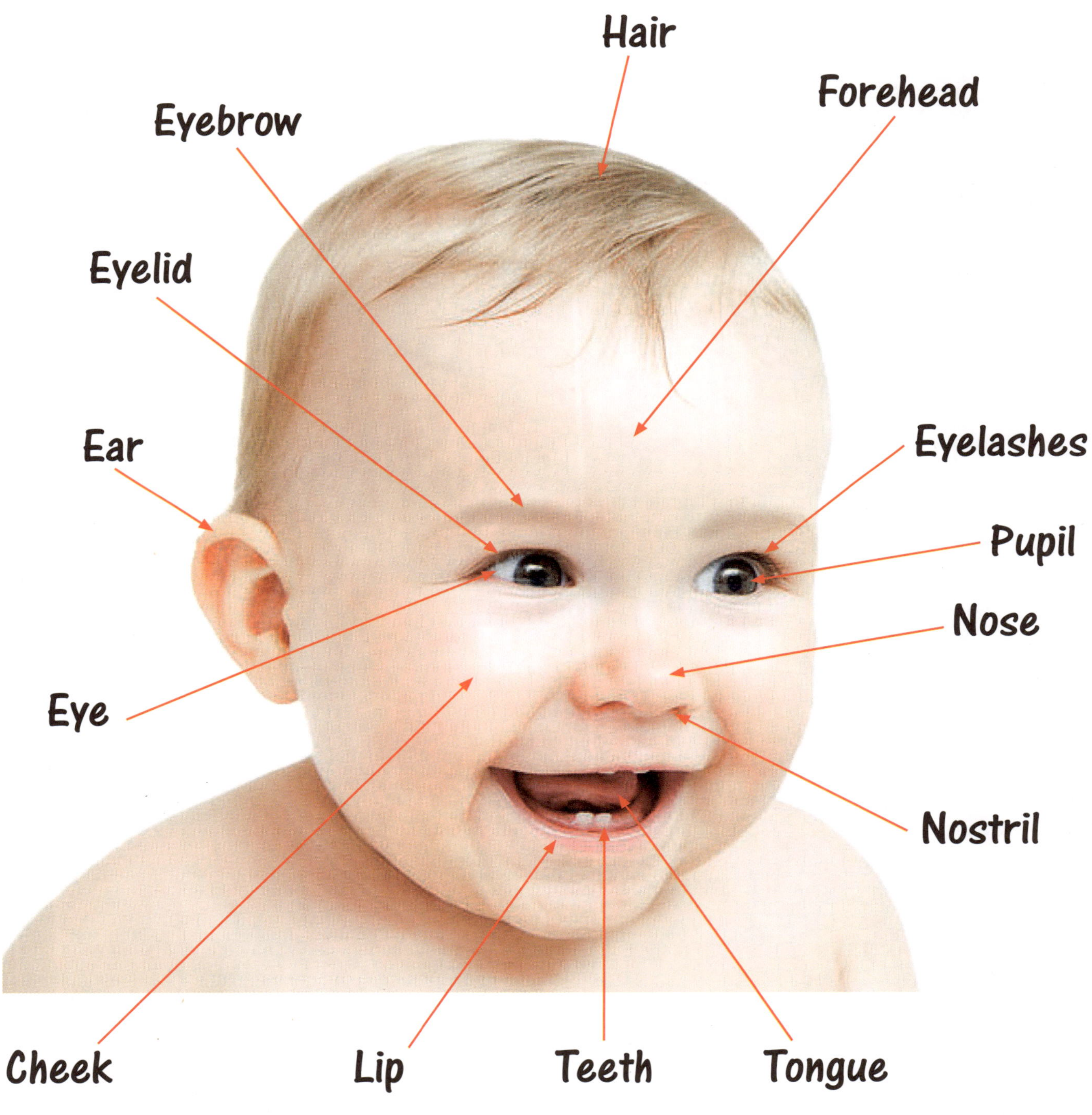
Hair
Eyebrow
Forehead
Eyelid
Ear
Eyelashes
Pupil
Nose
Eye
Nostril
Cheek
Lip
Teeth
Tongue

My Hand

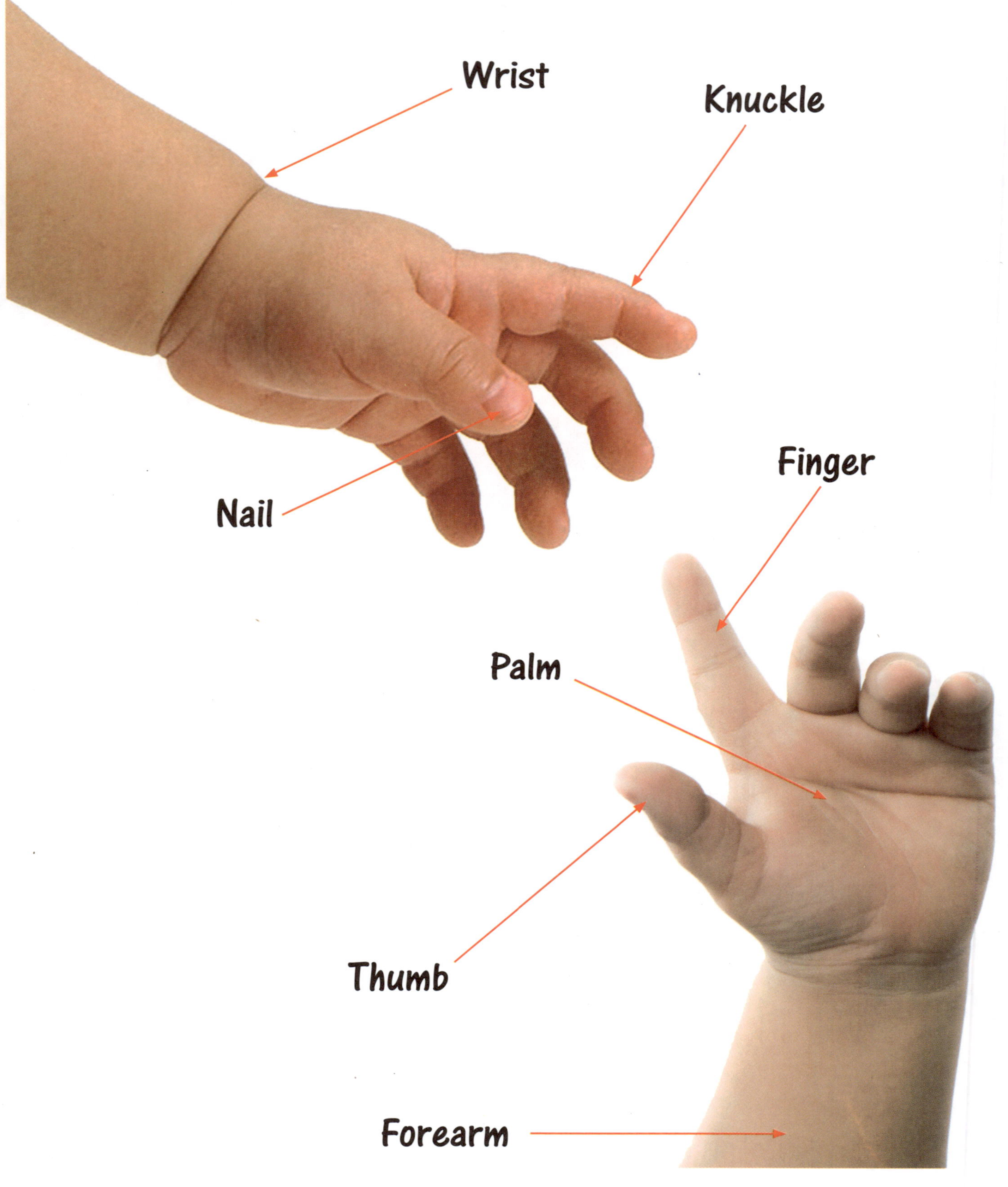

My Body

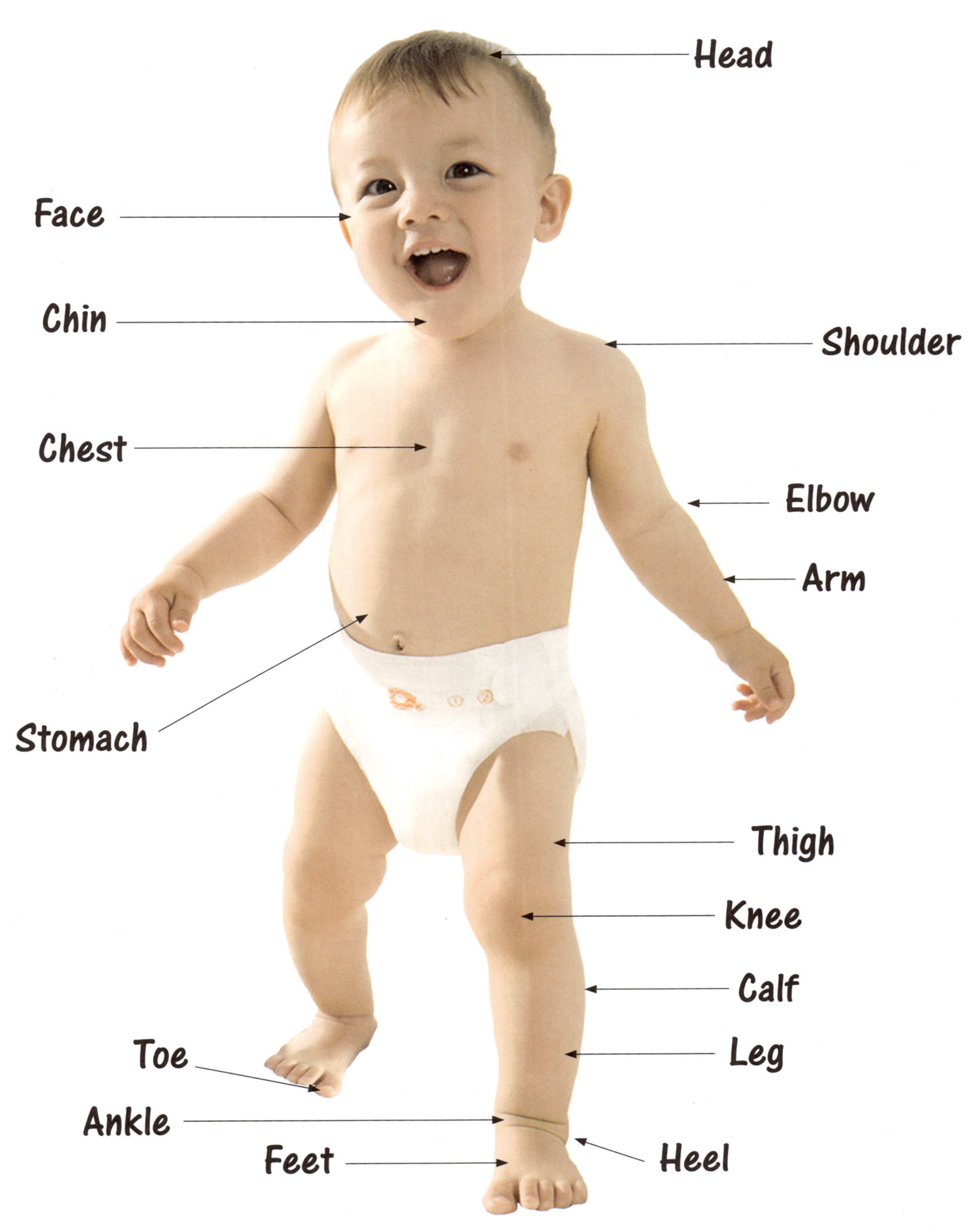
Head
Face
Chin
Shoulder
Chest
Elbow
Arm
Stomach
Thigh
Knee
Calf
Toe
Leg
Ankle
Feet
Heel

Words' List

A

Accessories 54
Action Words 56
Activity Toy 35
Aeroplane 2, 18
Air Conditioner 39
Air Transport 18
Alphabets 2
Ambulance 16
Ankle 61
Apple 14
Appliances 42
Apron 42
Archery 50
Architect 32
Arm 61
Aster 53
Astronaut 32

B

Baby Animals 30
Baby Platypus 31
Back 11
Bag 54
Baker 32
Ball 34
Banana 7
Bangles 54
Baseball 51
Basketball 51
Bathroom 40
Bathtub 37
Bear Cub 31
Bear 26
Bed 39
Bedroom 39
Beetroot 12
Belt 54
Bib 37
Big 10
Bike 35
Bird's House 46
Birds 22
Birthday Cap 7
Bitter Gourd 12
Black 6, 11
Blackberry 15
Blanket 36
Blazer 48
Blending 58
Blimp 18
Blocks 35
Blue 6
Bluebell 53
Boar 27
Board 45
Boat 19
Book 7
Books 44
Booties 37
Bottle 36
Bowl 37
Boxing 50
Bread 21
Broccoli 12
Brown 6
Bucket 40
Buffalo 25
Bull 25
Burger 20
Bus 2

C

Cabbage 12
Calf 30, 61
Camel 25
Canna 53
Capsicum 12
Car 2
Caravan 17
Cardigan 49
Cargo Ship 19
Carnation 52
Carpet 38
Carrot 12
Cat 24
Catfish 28
Cauliflower 13
Centre Table 38
Chair 38
Chalk 44
Cheek 59
Cheese 21
Cheetah 27
Chef 32
Cherry 15
Chest 61
Chicken 20
Chimney 43
Chimpanzee 27
Chin 61
Chopping Board 43
Circle 7
Clean 10
Clock 38
Close 11
Clothes 48
Clownfish 28
Cold 10
Colour 6
Colt 31
Computer 44
Cone 7
Cookies 20
Cooking Pan 42
Cooking Pot 42
Coral 29
Coriander 12
Cornflakes 20
Cow 25
Crab 29
Crane 22
Crescent 7
Crib 37
Cricket 50
Crocodile 26
Crow 22
Crying 58
Cucumber 13
Curtain 38
Custard Apple 14
Cycle 17
Cycling 51
Cygnet 31

D

Daffodil 52
Dahlia 52
Dancing 56
Deer 27
Desk 44
Diamond 7
Diaper 36
Dice 2
Digger 16
Dirty 10
Doctor Set 34
Doctor 32
Dog 24
Doll 34
Dolphin 29
Donkey 25
Doormat 39
Dove 23
Dragonfish 28
Drawer 39
Dressing Table 39
Drinking 57
Drum 34
Duck 22, 24
Dustbin 46
Duster 44

E

Eagle 22
Ear 59
Earrings 54
Easel 44
Eating 56
Eel 29
Eggplant 13
Eggs 7
Eight 8
Eighteen 9
Elbow 61
Electrician 32
Elephant Calf 30
Elephant 2
Eleven 9
Eraser 45
Exhaust Fan 41
Eye 59
Eyebrow 59
Eyelashes 59
Eyelid 59

F

Face 61
Fan 39
Farm Animals 25
Farmer 33
Fast 10
Fawn 31
Feeling Excited 58
Feeling Surprised 58
Feet 61
Ferry 19
Fifteen 9
Finger 60
Fire Truck 17
Firefighter 32
Fireplace 38
Fish 2
Fishing Boat 19
Five 8
Flamingo 22
Flowerpot 38
Flowers 52
Foal 30
Food 37
Foods 20
Football 7
Forearm 60
Forehead 59
Four 8
Fourteen 9
Fox 26

Frock 48
Front 11
Frowning 58
Fruit Salad 21
Fruits 14
Frying Pan 42

G
Garden Fork 46
Garden Hose 47
Garden 46
Gardener 33
Gas Lighter 43
Gas Stove 42
Ginger 13
Giraffe 27
Glass 43
Globe 44
Glue 44
Go Cart 17
Goat 3
Goldfish 25
Golf 50
Goose 23, 25
Gorilla 27
Gown 48
Grapes 15
Grass 47
Grey 6
Green Chilli 13
Green House 47
Green Pea 13
Green 6
Guava 15
Guinea Pig 24

H
Hair Band 54
Hair Brush 40
Hair Clip 54
Hair Dryer 41
Hair Oil 41
Hair 59
Hamster 24
Hanging Light 38
Happy 11
Hard 11
Hare 27
Hatchling 31
Hear 55
Heart 7
Heel 61
Helicopter 18
Hen 3, 25
Hibiscus 52
Hiding 57
High Chair 36
Highlighter 45
Hippo 26
Honeycomb 47
Horse Racing 51
Horse 24
Hot Air Balloon 18
Hot 10
Hotdog 20
Hyena 26

I
Ice Cream 3

J
Jacket 49
Jar 43
Jasmine 53
Jeans 49
Jeep 17
Jellyfish 29
Jet Ski 19
Jet 3
Joey 31
Jumping 56

K
Kangaroo 26
Karate 51
Kayak 19
Key Chain 54
Kid 31
Kingfisher 23
Kit 30
Kitchen 42
Kitchen Set 35
Kite 3
Kitten 31
Kiwi 14
Knee 61
Knuckle 60

L
Lady Finger 13
Lamb 30
Lamp 44
Land Transports 16
Laughing 57
Laundry Hamper 41
Leaf 7
Leaves 47
Leg 61
Lemon 3
Lily 52
Lion Cub 30
Lion 27
Lip 59
Listening 57
Living Room 38
Llama 25
Lobster 29
Long Coat 49
Loofah 41
Lorry 17
Lotus 53

M
Mango 14
Marigold 53
Marker 45
Microwave 42
Milk 21
Mirror 41
Mixer Grinder 42
Monkey 27
Motorcycle 16
Mouth Freshener 41
Muffin 20
Mug 41
Mushroom 4
Muskmelon 15
My Body 61
My Face 59
My Hand 60
My Senses 55

N
Nail 60
Nest 4
New 10
Nightgown 49
Nightingale 23
Nine 8
Nineteen 9
Nose 59
Nostril 59
Notebook 44
Notice Board 45
Numbers 8
Nurse 33

O
Objects 36
Octopus 29
Oil Tanker 17
Old 10
Omelette 21
One 8
Onion 12
Open 11
Opposites 10
Orange 6, 14
Orchid 52
Ostrich 23
Otter 28
Oval 7
Owl 4
Owlet 31

P
Pacifier 37
Palm 60
Panda 27
Pansy 52
Papaya 15
Parachute 18
Parrot 4
Peach 15
Peacock 22
Pear 14
Pen Holder 45
Pen 45
Pencil 45
Penguin 23
People at Work 32
Periwinkle 53
Pet Animals 24
Petunia 52
Phone 38
Photographer 33
Piano 35
Pick 56
Pig 25
Pigeon 23
Piglet 30
Pillow 37
Pilot 32
Pineapple 15
Pink 6
Pizza 20
Plate 43
Play Mat 37
Playing 56
Plum 14
Plumber 33
Polar Bear Cub 31
Polar Bear 27
Police Car 35
Police Officer 33
Pomegranate 14
Potato 13
Potty Seat 37
Pram 36
Praying 56
Pressure Cooker 42
Pulling 57

Pumpkin 12
Pupil 59
Puppy 30
Pushing 57

Q

Queue 4
Quilt 39

R

Rabbit 24
Raincoat 49
Rattle 34
Read 57
Rectangle 7
Red 6
Refrigerator 43
Remote Car 35
Reporter 33
Rhinoceros 26
Rice 21
Ring Donut 34
Ring 54
Robin 22
Robot 34
Rocket 18
Rocking Horse 35
Rolling Pin 42
Rooster 25
Rose 4
Rugby 51
Ruler 45

S

Sad 11
Sandwich 7, 21
Sausage 21
School Bag 45
School 44
Scientist 33
Scooter 17
Sea Aeroplane 18
Sea Animals 28
Sea Horse 29
Sea Turtle 28
Seal 29
See 55
Seven 8
Seventeen 9
Shampoo 41
Shapes 7
Shark 29
Shower 40
Sheep 25
Shells 28
Ship 5
Shirt 48
Shorts 48
Shoulder 61
Shrimp 28
Sipper 36
Sit 11
Six 8
Sixteen 9
Skateboard 16
Skiing 50
Skirt 48
Sleeping 58
Slow 10
Small 10
Smell 55
Smiling 56
Snowmobile 17
Soap Dish 40
Soap 40
Soccer 51
Socks 37
Sofa 38
Soft 11
Soldier 33
Soup 20
Sour 10
Spade 46
Sparrow 22
Spatula 43
Spinach 13
Sponge 40
Spoon 37
Sports 50
Sprinkle 46
Square 7
Squid 29
Stand 11
Star Fruit 14
Star 7
Starfish 7
Stingray 28
Stomach 61
Strawberry 14
Stroller 36
Submarine 19
Sunflower 52
Sunglasses 54
Swan 23
Sweatshirt 49
Sweet 10

T

Table Lamp 39
Table Tennis 50
Talking 58
Taste 55
Taxi 17
Teacher 32
Teddy 35
Teeth 59
Teether 36
Television 38
Ten 8
Tennis 51
Thigh 61
Thinking 56
Thirteen 9
Three 8
Thumb 60
Tiger Cub 31
Tiger 26
Tissue Paper 41
Toaster 42
Toe 61
Toilet Paper 40
Toilet Pot 40
Tomato 12
Tongue 59
Toothbrush 41
Toothpaste 40
Touch 55
Tow 17
Towel 41
Toy Cycle 34
Toy Train 34
Toys 34
Tractor 17
Train 5
Tray 43
Tree 46
Triangle 7
Trowel 46
Truck 16
T-Shirt 49
Tulip 53
Turnip 13
Turtle 24
Twelve 9
Twenty 9
Two 8

U

Umbrella 5

V

Vegetables 5
Violet 6
Volleyball 50
Vulture 22

W

Wallet 54
Walrus 29
Wardrobe 39
Washbasin 40, 43
Washing Machine 40
Washing Powder 40
Watch 5
Water Can 46
Water Lily 53
Water Transports 19
Watermelon 15
Weight Lifting 50
Whale 28
Wheelbarrow 46
Whisk 43
White 6
Wild Animals 26
Wolf 26
Woodpecker 23
Wrist 60
Writing 58

X

X Mas Tree 5

Y

Yacht 5, 19
Yawning 58
Yellow 6
Yogurt 21

Z

Zebra 27
Zip 5